The LIBRARY of LANDFORMS™

PENINSULAS

Isaac Nadeau

The Rosen Publishing Group's

PowerKids Press™

New York

To Papa and the old days

Published in 2006 by The Rosen Publishing Group, Inc.
29 East 21st Street, New York, NY 10010

First Edition

Editor: Rachel O'Connor
Book Design: Elana Davidian

Photo Credits: Cover © Robert Frerck/Getty Images; p. 4 (top left) © Image Makers/Getty Images; p. 4 (bottom) © Jeremy Horner/Corbis;
p. 4 (top right) © Paul A. Souders/Corbis; p. 7 © W. E. Garrett/National Geographic Image Collection; p. 8 (top) © Martha
Holmes/naturepl.com; p. 8 (bottom left) © Dean Conger/National Geographic Image Collection; pp. 8 (bottom right), 12 Digital Image ©
1996 Corbis, original image courtesy of NASA/Corbis; p. 11 (top) © Eric L. Wheater/Lonely Planet Images; p. 11 (bottom left) © 2002
GeoAtlas; p. 11 (bottom right) © Marc Muench/Corbis; pp. 12 (inset) © Jurgen Freund/naturepl.com; p. 15 © Jim Sugar/Corbis;
p. 16 © NASA/National Geographic Image Collection; p. 16 (inset) © James Randklev/Corbis; p. 19 (top) © Kevin Fleming/Corbis;
p. 19 (bottom left) © Adrian Davies/naturepl.com; p. 19 (bottom right) © Joe McDonald/Corbis; p. 20 © James L. Amos/Corbis.

Library of Congress Cataloging-in-Publication Data

Nadeau, Isaac.
  Peninsulas / Isaac Nadeau.— 1st ed.
    p. cm. — (The library of landforms)
  Includes index.
  ISBN 1-4042-3125-0 (lib. bdg.)
  1. Peninsulas—Juvenile literature. I. Title. II. Series.

GB454.P46N33 2006
551.41—dc22

                                           2004025429

Manufactured in the United States of America

CONTENTS

Top Left: Here you have a bird's-eye view of Italy, a peninsula, as seen from the spaceship *Columbia*. *Top Right*: Humpback whales can survive the cold temperatures of the waters off the Antarctic Peninsula. *Bottom*: The city of Salvador is located on a peninsula on the Atlantic coast. With a population of around 2.5 million people, Salvador is Brazil's third-largest city.

What Is a Peninsula?

A peninsula is a narrow body of land that is surrounded on three sides by water. The word "peninsula" comes from the Latin words for "almost" and "island." An island is a body of land surrounded on all sides by water. If you look at a map of the world, you will be able to find peninsulas on every **continent**. Peninsulas can be large, such as the country of Italy in the Mediterranean Sea or the Baja Peninsula in the Pacific Ocean. Larger peninsulas may have millions of people living on them. For example, about 68 million people live on the Korean Peninsula on the east coast of Asia. There are also thousands of small, unnamed peninsulas in oceans, lakes, and rivers all over the world. Many of these peninsulas are **uninhabited**.

Peninsulas can be home to many different types of animals and plants. For example, around 480 different **species** of birds live on the Florida Peninsula. However, there are other peninsulas with few living things on them. Few plants or animals can **survive** the cold and ice of the Antarctic Peninsula on the southern continent of Antarctica. Here winter **temperatures** can reach -70°F (-57°C).

How Peninsulas Are Formed

Every peninsula is connected to a larger body of land. For example, Cape Cod is connected to the eastern coast of North America, the Alaska Peninsula is connected to the western coast, and the Florida Peninsula is connected to the southeastern coast. However, these peninsulas were formed in very different ways. To learn how peninsulas are formed, **geologists** look at the rock from which the peninsulas are made. Some peninsulas, such as the Alaska Peninsula, are formed of **volcanic** rock. When a volcano **erupts**, **magma**, or melted rock, flows out of Earth's surface. When this magma cools, it hardens, adding new rock to the land where the volcano erupted. If the volcano is close to the water, the new rock can form a peninsula. Some peninsulas, such as the Antarctic Peninsula, are made up of **sedimentary rock** that was **deposited** by **glaciers** or streams. Others are made entirely of sand. Some peninsulas, such as the Florida Peninsula, have taken millions of years to form. Others, including many of the small **barrier** peninsulas off the eastern coast of North America, can be formed and destroyed in a matter of days.

Peninsulas can be formed by volcanic mountains. Here steam rises from the volcano, Mount Martin, which is found in Katmai National Park at the tip of the Alaska Peninsula.

4.5 BILLION YEARS AGO:	ABOUT 3.8 BILLION YEARS AGO:	500 MILLION – 200 MILLION YEARS AGO:	ABOUT 200 MILLION YEARS AGO:	ABOUT 23,000 YEARS AGO:	TODAY:
Earth is formed.	The liquid rock that makes up the surface of Earth begins to cool and harden, forming the first continents. The first oceans also form. As a result the first peninsulas on Earth were probably formed around this time, as well.	All of the continents on Earth move together to form a supercontinent called Pangaea.	Pangaea begins to break up, and North America breaks away from North Africa, as the ocean flows between them. This is when the Florida Peninsula was formed.	Cape Cod Peninsula is formed by till that washed out of a melting glacier.	The many peninsulas that are found on Earth today are in a constant state of change. Deposition causes some to become larger. Erosion and changes in sea levels are causing some to disappear altogether.

Top: The Antarctic Peninsula is on the western coast of Antarctica. Most of Antarctica is covered with ice, and the peninsula is the warmest area. Above-freezing temperatures are common, which can result in ice floes from melting glaciers, as shown here. *Bottom*: Cape Cod is a peninsula formed by glaciers. On the left you can see one of its shorelines up close, and on the right is the whole of Cape Cod from above.

Peninsulas Formed by Glaciers

Cape Cod, on the eastern coast of North America, is an example of a peninsula formed by rocks left behind by glaciers. A glacier is a large sheet of ice that moves slowly across the land during cold periods known as ice ages. Glaciers carry many rocks of different sizes. The rocks can range from tiny bits of clay to **boulders** the size of houses. The rocks in a glacier are called till. When temperatures increase, water from a melting glacier washes the till away. In some areas this till forms large hills. A peninsula forms when these hills are formed close to water but are still connected to the mainland.

On Cape Cod, much of the peninsula is made up of sand, gravel, and boulders that were once carried by a large glacier that covered much of northern North America. About 23,000 years ago, this glacier covered the area that is now Cape Cod. When temperatures rose and the glacier melted, a lot of till was left behind. This till formed most of the peninsula that is Cape Cod.

On ocean coasts the water rises and falls with the tides once or twice a day. Many peninsulas can only be seen at low tide. When the tide comes in, the peninsula may disappear altogether under the water.

Volcanic Peninsulas and Rift Peninsulas

The Kalaupapa Peninsula, off the island of Molokai in the Hawaiian Islands, is an example of a volcanic peninsula. Volcanic peninsulas are formed by the **molten** rock, or lava, that flows out of active volcanoes. An active volcano is a volcano that has erupted and is likely to erupt again. About 230,000 years ago, a volcano erupted on the ocean floor, north of Molokai. The lava that flowed out of the volcano cooled and hardened in the water. As the volcano continued to erupt, the lava piled higher until it reached above the surface of the ocean, forming an island. More lava poured out of the volcano, adding more rock to the island. In time this rock connected to Molokai, forming the Kalaupapa Peninsula.

Rift peninsulas form when a piece of land slowly rips away from a continent. Rifting is a result of **plate tectonics**, which is the movement of the plates that make up Earth's crust. As the plates move against or apart from one another, parts of Earth's crust can be ripped apart, like pieces of paper. The Arabian Peninsula is an example of a peninsula formed in this way.

The Kalaupapa Peninsula is about 2 miles (3.2 km) long. The most popular way to visit the Kalaupapa Peninsula is to travel from Molokai by mule! Here you can see a group doing just that. *Bottom Left*: This map shows the Baja Peninsula. The Baja Peninsula is a rift peninsula. About 10 million years ago, Baja was part of the Mexican mainland. Over the last 5 to 10 million years, the Baja Peninsula has been pulling slowly away from Mexico. *Bottom Right*: This is a rocky shore on the Baja Peninsula.

Here you can see the Mississippi River delta as it is seen from space. This picture was taken from the spaceship *Challenger*. *Inset*: This aerial view of a river in Queensland, Australia, shows sediment forming at the river's mouth, where it enters the ocean.

Delta Peninsulas

Many peninsulas are formed where large rivers flow into the ocean. As rivers flow downhill, they pick up and carry large amounts of sand and rock. This sand and rock is called sediment. Sediment can be carried hundreds of miles (km) until it reaches the sea. At the mouth of the river, or the place where the river meets the ocean, the water slows down causing the sediment to be deposited. The largest sediment is deposited first because it is the heaviest and sinks to the bottom as soon as the river slows down. Lighter sediment, such as sand and clay, is carried farther out into the ocean. Wherever a river flows into the ocean, much sediment can be found on the ocean floor near the river's mouth. In some cases so much sediment is deposited that a pile rises above the surface of the ocean, and a peninsula is formed. The sediment traveling down the Mississippi River toward the Gulf of Mexico formed the peninsula called the Mississippi Delta. Even today the Mississippi River is washing thousands of tons (t) of sediment down to the sea every year. As this sediment is deposited, the Mississippi Delta is growing larger.[1]

Sand Spits

Sand spits are a type of small peninsula. They are formed by sand and are the result of the movement of ocean waves washing against a coast. Waves help erode rock by crashing against cliffs and breaking off small pieces. Rocks, shells, and sand are tossed about in the constant motion of the waves, causing the bits of sand to become smaller and smaller. This sand is dragged into the ocean and carried in the direction the waves are moving. As the waves move along the shore, they deposit much of this sand. In some cases the waves form long, narrow lengths of sand that are surrounded on three sides by water and connected to the shore on one side. These small peninsulas are called sand spits.

Sand spits are always changing shape because of the forces of erosion and deposition. Sometimes waves will erode the place where a sand spit connects to the mainland, forming an island. Many of the barrier islands off the coast of North Carolina were formed in this way. In some cases a storm, such as a **hurricane**, can change the size and shape of sand spits in a very short time.

Here you can see a sand spit that has been formed at the mouth of the Columbia River, where it meets the Pacific Ocean.

This is a picture of Florida taken from space. More than 15 million people live on the Florida Peninsula. *Inset*: This beach on Jupiter Island, Florida, has banks of limestone rocks on its shoreline.

THE FLORIDA PENINSULA

The Florida Peninsula is the most famous peninsula in the United States. The rocks that make it up are thousands of feet (m) thick. The rocks at the bottom were deposited first and are the oldest. Many geologists believe that the oldest rocks on the Florida Peninsula were once part of what is now northwestern Africa. About 200 million years ago, Africa and North America were joined. Because of rifting, North America and Africa began to separate. Water from the surrounding oceans began to flow between them, and the Atlantic Ocean was formed. Much of the ancient rock of Florida was covered by the ocean. Many tiny animals lived and died in this ocean. For millions of years, their shells and bones sank to the bottom of the sea, falling on the ancient rocks. These shells and bones piled higher and higher, and **eventually** they were pressed together into a type of sedimentary rock called limestone. Over millions of years, this limestone grew to be thousands of feet (m) thick in some places. When the ocean level dropped, much of this stone was exposed at the surface, forming part of what is now the Florida Peninsula.

LIFE ON THE FLORIDA PENINSULA

More than 2,000 species of plants grow in the Everglades. Trees such as the mahogany and the cypress grow there. The Everglades is also home to about 350 species of birds, including the wood stork, which hunts by feeling for fish in the water with its bill.

Around the world plants and animals make their homes on peninsulas. For example, the Florida Peninsula is a place with a great range of living things. This is because it is warm year-round, has a lot of ocean shoreline, and has many streams and lakes.

At the southern end of the peninsula, the Florida Everglades is especially rich with life. The Everglades is a large, low-lying area. Because the peninsula is so flat and low, the Everglades is flooded with water every spring and summer during the rains of the wet season. The Everglades slowly gets drier in the winter. During the wet season, tiny plants called algae and tiny animals called plankton grow quickly. Plankton and algae are food for many larger animals. These animals include salamanders, frogs, snails, and mosquitoes. Salamanders and frogs eat mostly bugs. The smallest frog in North America, the little grass frog, lives in the Everglades. Other animals include alligators and snakes.

Top: The Everglades is a great place to watch birds. In this photograph you can see herons and egrets as they fly around their nesting area in the Everglades. *Bottom Left*: The largest reptile in North America, the American alligator, lives in the Everglades. It feeds on fish, turtles, birds, and other animals that it catches in its huge, toothy mouth. *Bottom Right*: The rat snakes found in the Everglades usually measure about 5 feet (1.5 m) long.

The Keweenaw Peninsula is located in Michigan. It is surrounded on three sides by Lake Superior. Here the waves of Lake Superior can be seen crashing against the shores of the peninsula.

Peninsulas over Time

Peninsulas are always changing. They are changed by wind, waves, and rain. They are changed by the addition of new sediment from streams. They may become wider, narrower, longer, or shorter. They may become islands or disappear underwater altogether. There are many ways in which a peninsula can change shape. For example, the southern shore of Cape Cod is being slowly washed away by waves. About 5 acres (2 ha) per year of Cape Cod's southern shore are washed away by waves. If this continues for a long time, the whole peninsula of Cape Cod will eventually be washed away.

Other peninsulas change as the sea level rises or falls. Many scientists think that temperatures on Earth are getting warmer. As this occurs Earth's glaciers are melting, and the water is flowing into the oceans. As the oceans rise with this water, many peninsulas will become covered in more water. As a result the peninsulas will get smaller. Many peninsulas, such as Florida and the Mississippi Delta, are very close to sea level now. A rise of a few feet (m) would cover much of these peninsulas in water.

People and Peninsulas

Many peninsulas are popular places for people to live because they are close to the water. If you look at a map of the cities of the world, you will find that many of them are on peninsulas. People live near water for many reasons. We need water to survive. Water also helps us carry goods from one place to another. People also travel on water using boats. Water is where fish and other foods can be found. Surrounded on three sides by water, peninsulas are therefore often good places to live. Peninsulas are also good places to visit. This is because of the beauty of the shore and the many fun things to do on the water. For example, thousands of people travel to Florida each winter to enjoy the warm temperatures and the beaches. Peninsulas are also important **landmarks.** Many peninsulas have lighthouses on them to show ships that they are getting close to land.

Every peninsula is special in its own way, wherever it is found. One of the best ways to learn about peninsulas is to look for one near where you live and explore it yourself.

Glossary

barrier (BAR-ee-er) Having to do with something that blocks something else from passing.

boulders (BOL-durz) Very large rocks that are usually round.

continent (KON-tih-nent) One of Earth's seven large landmasses.

deposited (dih-PAH-zuht-ed) Left behind.

erupts (ih-RUPTS) Bursts out of something.

eventually (ih-VEN-chuh-wel-ee) At some point.

geologists (jee-AH-luh-jists) Scientists who study the form of Earth.

glaciers (GLAY-shurz) Large masses of ice that move down a mountain or along a valley.

hurricane (HUR-ih-kayn) A storm with strong winds and heavy rain.

landmarks (LAND-marks) Buildings or places that are worthy of notice.

magma (MAG-muh) A hot, liquid rock underneath Earth's surface.

molten (MOL-ten) Made liquid by heat.

plate tectonics (PLAYT tek-TAH-niks) The study of the moving pieces of Earth's crust.

sedimentary rock (seh-dih-MEN-teh-ree ROK) Layers of gravel, sand, silt, or mud that have been pressed together to form rock.

species (SPEE-sheez) A single kind of living thing. All people are one species.

survive (sur-VYV) To stay alive.

temperatures (TEM-pruh-cherz) How hot or cold something is.

uninhabited (un-in-HA-beh-tid) Not lived in.

volcanic (vol-KA-nik) Having to do with a volcano, which is an opening in Earth's surface that sometimes shoots up a hot liquid rock.

Index

A
Alaska Peninsula, 6
Antarctic Peninsula, 5

B
Baja Peninsula, 5
barrier islands, 14
barrier peninsula(s), 6

C
Cape Cod, 6, 9, 21

F
Florida Everglades, 18
Florida Peninsula, 5–6, 17, 18

G
glaciers, 6, 9, 21

I
Italy, 5

K
Kalaupapa Peninsula, 10
Korean Peninsula, 5

L
lava, 10

M
Mississippi Delta, 13, 21

P
plate tectonics, 10

R
rift peninsula, 10

S
sand spits, 14

T
till, 9

V
volcano, 6, 10

Web Sites

Due to the changing nature of Internet links, PowerKids Press has developed an online list of Web sites related to the subject of this book. This site is updated regularly. Please use this link to access the list:
www.powerkidslinks.com/liblan/peninsula/